What is Sand?

A Coloring Book by
The Georgia Mineral Society, Inc.

Written by Lori Carter

This edition published by:

The Georgia Mineral Society, Inc.
4138 Steve Reynolds Boulevard
Norcross, GA 30093-3059
www.gamineral.org

ISBN: 978-1-937617-04-2

What is Sand?

The Georgia Mineral Society, Inc.

What do you think sand is?

Sand can be tiny minerals and rocks

Did you know sand is not just tiny rocks?

Sand can contain bits of shells

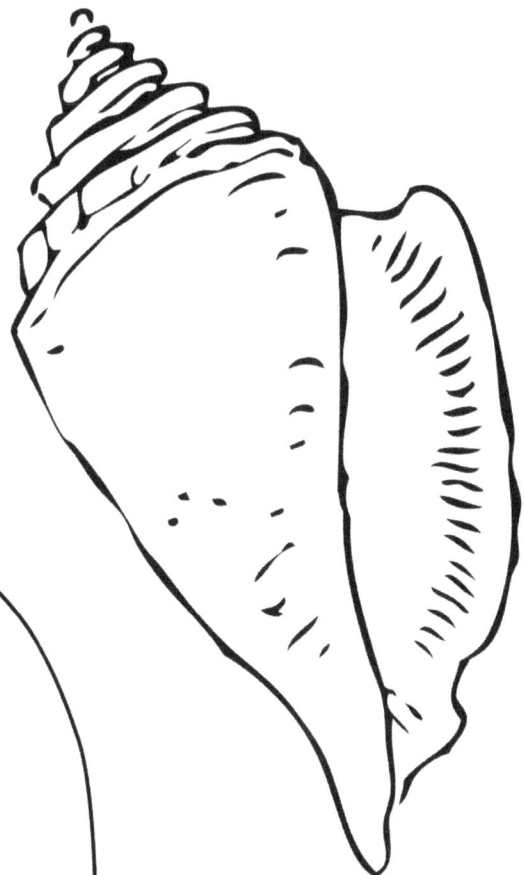

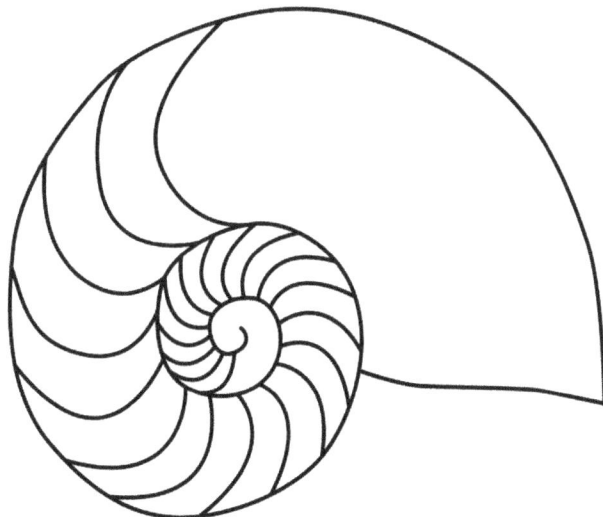

Sand can contain tiny fossils

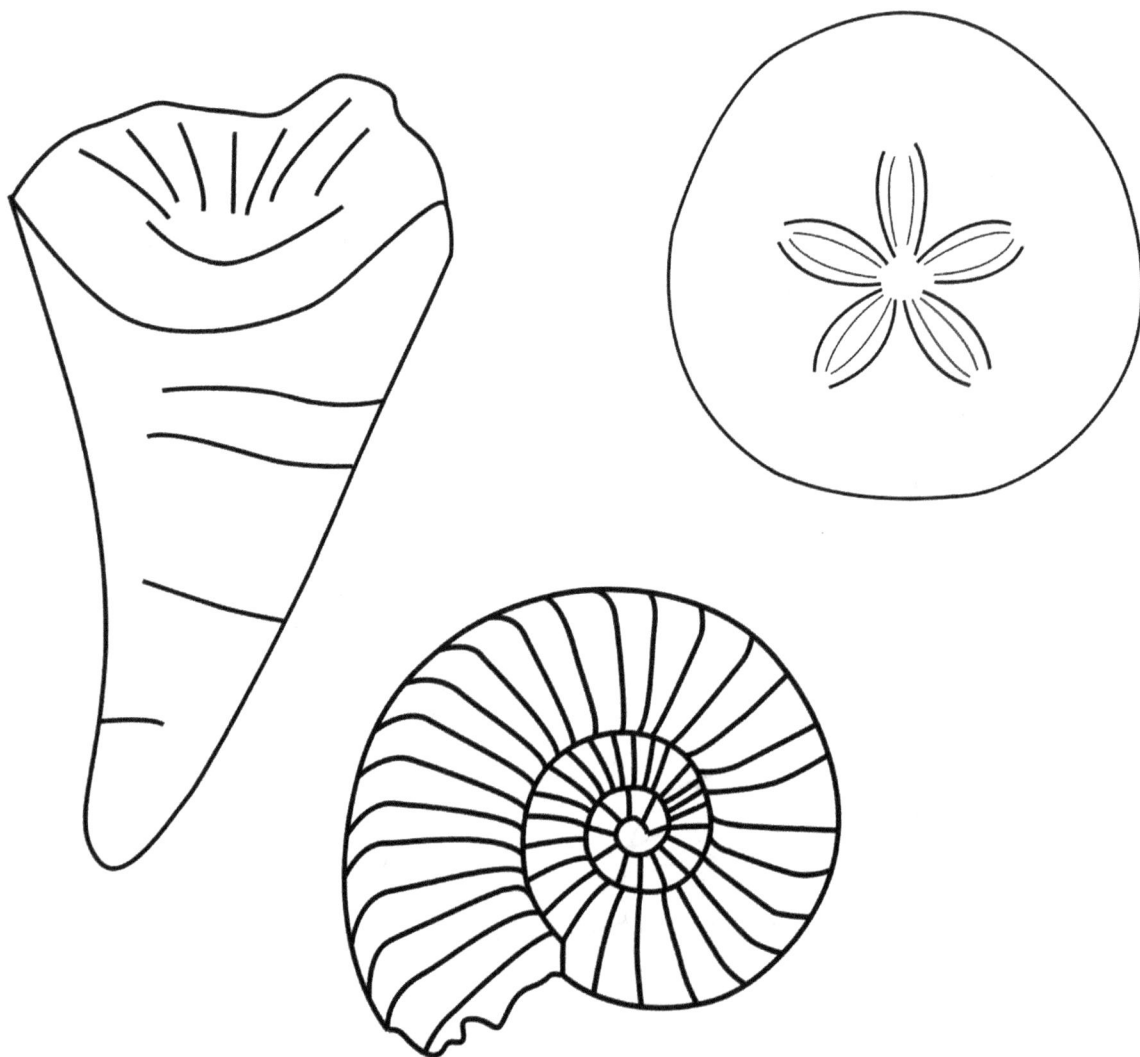

Sand can contain bits of glass

The Georgia Mineral Society, Inc.

Sand can be round things called ooliths

(oh-uh-liths)

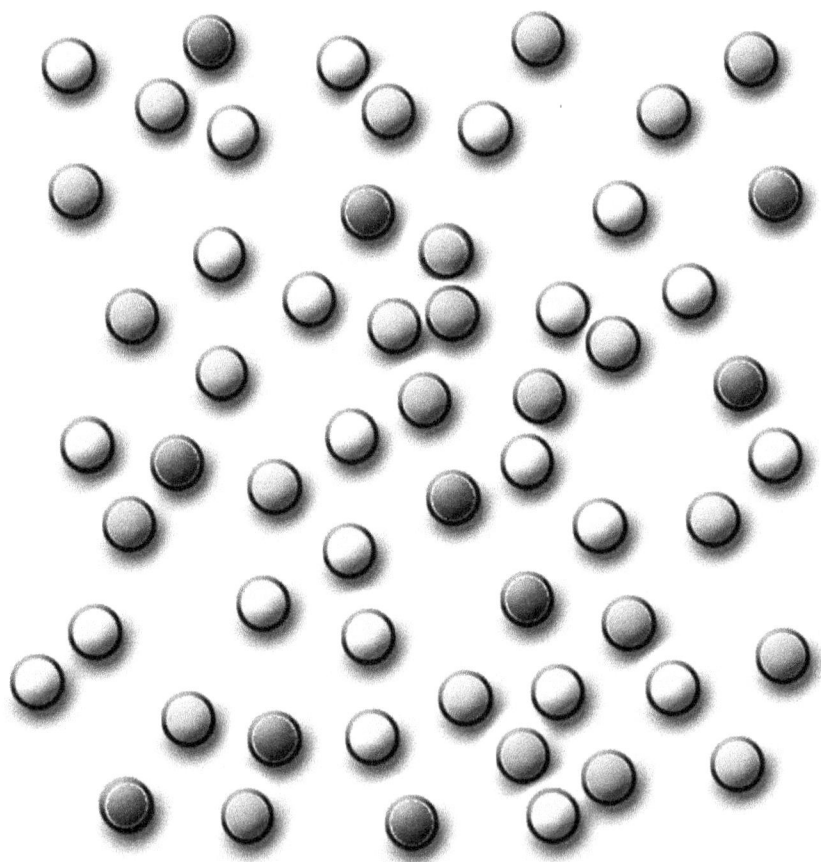

Sand can even look like little stars!

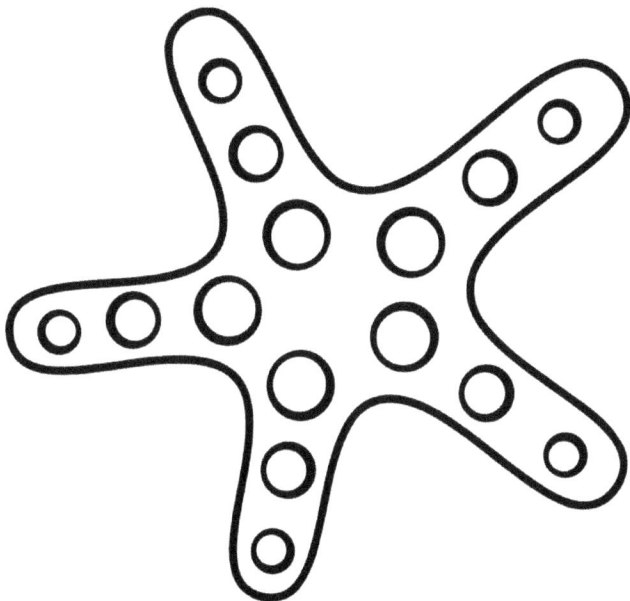

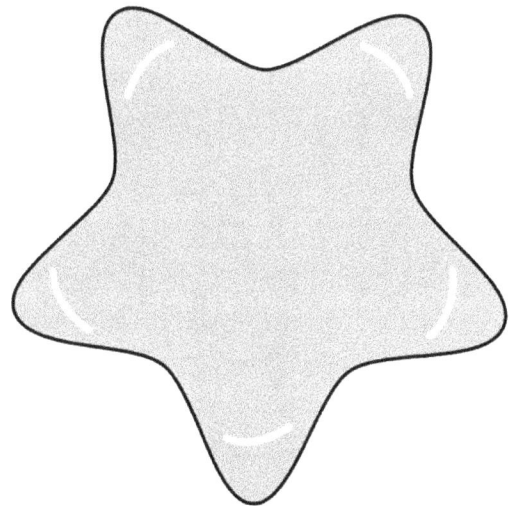

What color do you think sand is?

Sand can be brown

Did you know sand is not always brown?

Sand can be black

Sand can be white

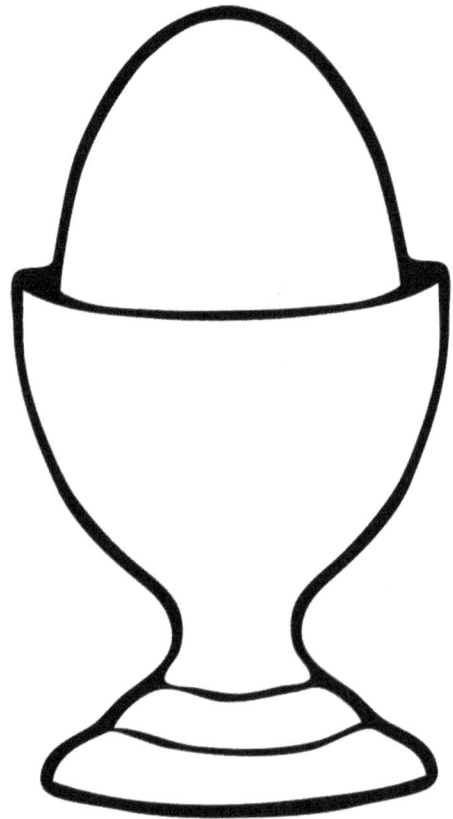

Sand can be green

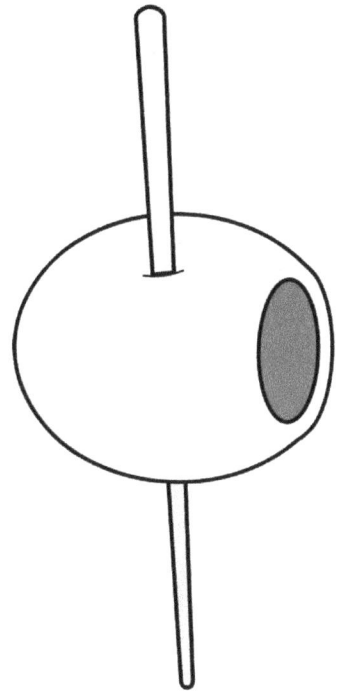

Sand can be orange

Sand can be red

Sand can be yellow

Sand can be many colors!

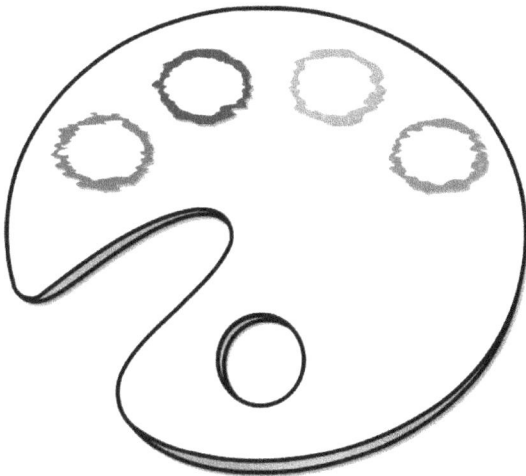

Where do you think you can find sand?

Sand can be found at the beach

Did you know sand is not always at a beach?

Sand can be found in a desert

Sand can be found at a lake

Sand can be found at a river

It might even be in your backyard!

Let's play a sand game!

Find these words about sand!

What can sand be made of?

rocks	fossils
shells	glass
ooliths	minerals

What color can sand be?

brown	red
black	yellow
white	orange
green	many

Where can sand be found?

beach	river
lake	desert
backyard	

The Georgia Mineral Society, Inc.

r o c k s x y r i v e r c y
z d p m h l d s g h u x e d
r e d y e f r t b q g l x m
d a n y l k p n r k l a u i
z j p a l f b u o o n k h n
k w d e s e r t w g h e j e
m l p z b n m y n o t w y r
a v j z h o k p w b r b m a
n w t o r a n g e g r a n l
y t u p z q j l y n f c l s
b g h m a f m a q e r k t i
o b e a c h e s r e q y z t
w k l d j t o s z r m a v w
j p g f q w e r t g y r f h
f o s s i l s m p b g d q i
e o o l i t h s l s t s r t
b x k d l p k c a l b q v e

r o c k s   r i v e r   y

      h   e

r e d   e   b   l   m

      l   r   a   i

      l   o   o   k   n

  d e s e r t w   e   e

                   n   r

m   b   a

a   a   l

n   o r a n g e   c   s

y   l   n   k

          a   e   y

  b e a c h   s   e   a   w

          s   r   r   h

               g   d   i

f o s s i l s

o o l i t h s   t

      k c a l b   e

Fun Facts About Sand!

There is a beach in Hawaii where the sand is green! The green is a mineral called "olivine".

A beach in Japan has sand that looks like little stars! The stars are the skeletons of tiny creatures called "foraminifera".

Next to a river in Georgia, the sand sparkles on a sunny day! It is full of a mineral called "mica" that makes it so shiny.

In many tropical places, the sand is white and pink and feels soft and smooth. It is made of tiny bits of shell and coral that have been rounded and smoothed by the ocean.

When metal ores are melted, glass can form on top of the molten metal. The glass is called "slag". Sometimes the slag is crushed into sand and used for concrete. Recycled, crushed glass sand can be used in concrete too.

The End